ASSERTIVE DISCIPLINE

SECONDARY RESOURCE MATERIALS WORKBOOK

Grades 7-12

Lee Canter

with

Marlene Canter

Canter and Associates, Inc.
Santa Monica, California

Printed in the United States of America

ISBN: 09608978-5-2

First Printing August 1984

Editorial Staff
    Kathy Winberry
    Barbara Schadlow

Art Director
    Kris Kegg

Consultants
    Bert Simmons
    Harriett Burt

Secondary students are more difficult to deal with today than ever before. They act out and defy authority in ways that students in the 50's and 60's would never have dared. In many cases educators cannot even turn to parents for help. Many parents are either unable or unwilling to cooperate with the school in disciplining their teen-age children. This leaves you, the educator, in a difficult position.

Assertive Discipline is a comprehensive program formulated to reverse behavior problems in the classroom from the moment the plan is implemented. This Workbook is designed to help meet the specific needs of secondary educators who are developing and carrying out an Assertive Discipline Plan.

The first section of this Workbook gives the individual teacher guidelines for establishing an effective Discipline Plan geared to situations that occur within the classroom.

The second section discusses the schoolwide plan and how to handle problems in the common areas of the school.

As you read through each section, keep in mind that the key to solving discipline problems is consistency. You, the educator, must never let up with your expectations and standards. Secondary students need to know that *every single time* they misbehave they will be provided with a consequence.

At the same time you provide negative consequences, you must balance your disciplinary actions with positive support of students' commendable and appropriate behavior. Secondary students, though they may appear aloof and disinterested, do like and need positive reinforcement — all people do.

One last point. Your Assertive Discipline Plan should be continually evaluated and analyzed. When it's not working, change it and get tougher. Never give up. If you remain determined in your efforts and attitude, you can solve the discipline problems you face, and then be able to go ahead and do what you do best — teach!

Introduction

Table of Contents

Introduction

Section One: How to Develop a Classroom Assertive Discipline Plan

Section Two: Classroom Reproducibles

Section Three: How to Develop a Schoolwide Assertive Discipline Plan

Section Four: Schoolwide Reproducibles

How to Develop a Classroom Assertive Discipline Plan

Establish Behavior Rules for Your Classroom

The goal in developing a classroom Assertive Discipline Plan is to have a fair and consistent way in which to deal with all students who misbehave, thereby creating an atmosphere conducive to teaching and allowing more time on-task for learning.

First, it is vital that you determine the specific behaviors you will require of your students. These behaviors will serve as the rules for your classroom. Take time to think carefully about the behaviors you really need, then choose a maximum of five to be used as the rules for your classroom.

When determining rules be sure that they are observable. Rules such as "Raise hand to speak" are observable. Rules such as "Show respect" are too vague and not observable.

The key rule that must be in effect at all times is: Students must follow your directions the *first time* they are given.

Most secondary teachers use a combination of classroom management rules and specific rules geared to the subject area taught.

Here are examples of observable classroom management rules that secondary teachers use:
- Follow directions the first time they are given.
- Be in class and seated when bell rings.
- Bring books, notebooks, pen and pencils to class.
- Raise hand to be recognized before speaking.
- Hand in all assignments on time.
- Keep hands, feet and objects to yourself.

Here are some examples of specific subject area rules that secondary teachers use:
- Clean up all lab work.
- Be in place, in gym clothes, five minutes after the bell rings.
- Students will remain in seat and do individual work in Study Hall.
- Follow all safety rules.
- Absolutely no talking during classroom study time.
- Homework must be handed in every day.

When determining your rules, *remember that good classroom management begins with clearly defined standards. Your students must know your expectations in order for them to behave appropriately.*

Behavior Rules
for Your Classroom

List the rules for your classroom.

1.

2.

3.

4.

5.

Determine Disciplinary Consequences for Your Classroom

Choosing Your Disciplinary Consequences

Once your rules are established, you must determine the disciplinary consequences you will use for students who choose to misbehave. It is important that you take the time to carefully select consequences that are appropriate for your particular teaching situation. Use these guidelines to help you determine the disciplinary consequences you will use:

Choose consequences you are comfortable with. (For example, do not keep students after school if you are not comfortable staying after school.)

The consequences should be something the students do not like, but under no circumstances should they be physically or psychologically harmful to the students.

Choose a maximum of five consequences and list them in order of severity. (Some secondary teachers use only three consequences by deleting the warning and making the tougher consequences occur sooner.)

These consequences will become your discipline hierarchy. The number of times a student breaks a rule will determine the consequence the student will receive.

Typical disciplinary hierarchies are:

1st time student breaks rule:	Name on board	=	Warning
2nd time student breaks rule:	Name √	=	One detention after school
3rd time student breaks rule:	Name √√	=	Two detentions after school
4th time student breaks rule:	Name √√√	=	Two detentions after school, call parents
5th time student breaks rule:	Name √√√√	=	Two detentions after school, call parents, student sent to vice-principal

Remember, sending a student to principal or vice-principal should be near the end of the hierarchy.

1st time student breaks rule	=	45-minute campus clean-up
2nd time student breaks rule	=	One hour campus clean-up, call home
3rd time student breaks rule	=	In-school suspension, send to vice-principal

If you are not comfortable recording names or checks on the chalkboard, record them in a notebook or on a clipboard - any way you prefer. If students cannot see the checks you have recorded, you must tell them what consequences you have recorded. (For example, "Linda, Rule 2." or, "Linda, that's 3 checks.")

The disciplinary hierarchy should also include a severe clause. This clause should state that in the case of severe misbehavior such as fighting, vandalism, defying a teacher, or stopping the class from functioning, the discipline hierarchy no longer applies. Instead, an immediate consequence will occur. Some severe clause consequences are:

- Send the student to the dean of students
- Send the student to the principal
- The student receives in-school suspension (See page 48 for details.)

To assist you in developing a discipline hierarchy, here are some examples of consequences that secondary teachers use:

- Last one to leave classroom (student to remain 30 or 40 seconds in seat after the last student has left the classroom)
- Citation
- After-school detention
- Noontime detention
- Letter to parents
- Send to principal
- Campus clean-up
- In-school suspension
- Send to vice-principal
- Supervised lunch
- Send to another room
- Assigned a designated seat
- Demerits applied to citizenship grade
- Room clean-up

When determining your hierarchy, remember: It is not the severity but the inevitability of receiving the consequence that has impact!

Implementing Your Hierarchy

In order to effectively implement the discipline hierarchy that you have developed, follow these guidelines:

Every time a student breaks one of your rules, provide him or her with a disciplinary consequence.

The consequence should occur as soon as possible after the student misbehaves. (Be sure to notify the parent in advance before detaining a student after school.)

Stay calm when disciplining a student.

Make sure the students are given due process, i.e., they must know beforehand the consequences that will occur as a result of their misbehavior.

Parent contact should always be included towards the end of the hierarchy.

If sending a student to the principal is part of the hierarchy, it should be one of the last steps (except in cases of severe misbehavior).

At the end of each day or period, erase the names and checks so the students may begin with a clean slate the next day.

If your disciplinary consequences do not work, try "tougher" ones. If you have to use the same consequence with one student three times, it means the consequence is not working. For example, if Joe has received lunch detention three times in one week, your plan must be made tougher. You should drop down on the hierarchy or devise more severe consequences. "Joe, from now on, if there is one check next to your name, instead of lunch detention you will receive in-school suspension and I will call your parents." Be sure to inform the parents of your new plan.

When implementing your disciplinary hierarchy, remember that you need to be as "tough" as the kids are. When dealing with teen-agers many adults become frustrated and simply "give up." However, if you stick to your standards and follow through with your consequences consistently, most students will see that you really mean what you say and they will eventually change their behavior accordingly.

Disciplinary Consequences

List your disciplinary consequences.

First time student breaks a rule

Second time student breaks a rule

Third time student breaks a rule

Fourth time student breaks a rule

Fifth time student breaks a rule

Severe clause

Determine Positive Reinforcement for Your Classroom

The next step in developing an Assertive Discipline Plan is to determine how you will positively reinforce students who do behave. *Remember, the key to effective discipline is positive support of students' appropriate behavior. Negative consequences stop inappropriate behavior, but only positive consequences will change behavior.* Use these guidelines to help you decide how to respond to students who behave appropriately.

Establish responses with which you are comfortable.

Positive reinforcement should be something the students like.

Students should be informed of the positive reinforcement they will receive.

Positive responses should be provided as often as possible.

Plan ahead of time which specific appropriate behavior merits reinforcement.

Rewards should never be taken away as punishment.

Positively Reinforce Individual Students

The most effective reinforcer is verbal praise. In general, secondary students do not like to be singled out in front of their peers. Instead they should be praised after class or privately. However, students may be praised in front of their peers in an activity-based class, such as athletics, drama, or choir when students are working together on a joint project. Such praise encourages the group effort.

Non-specific praise creates praise "junkies." Verbal praise should be very specific. Non-specific praise such as "You did a good job today" or "Excellent work" is too vague. In order for students to repeat their good behavior they need to know specifically what they have done right! More appropriate would be comments such as "I like the way you contributed to the class discussion today," or "The work you handed in was exceptionally neat."

Try some of these other positive reinforcers for individual students:
- A positive note discreetly handed to a student, placed on his or her desk, or written on the second page of an assignment
- A positive note or call to parents
- A positive note mailed home addressed to student
- Skip a homework assignment (see page 31)
- Extra computer time
- Gift certificate (e.g., McDonald's)
- Discount at school store
- Free admission to school function
- Excused from one pop quiz
- Take a problem off test

Positively Reinforce the Entire Class

Marbles-in-a-jar or points are effective ways of recording classwide positive behavior. Select an activity that students can work for, then determine the number of marbles or points they must receive in a designated period of time (for example, the class will earn free radio time if twenty points are achieved in five periods). Use the following guidelines to set up a system using marbles or points to help you remain consistent.

Students earn up to five marbles or points per period for appropriate behavior.

The class earns the reward quickly, within one or two weeks.

Never take marbles out of the jar or points off of the board.

When the class earns one reward, begin again with a new goal.

Use peer pressure. (For example: No names on the board = three marbles or three point bonus.)

Give a direction, then reinforce with marbles or points.

Use competition among classes to encourage students to behave.

Here's how it's done:

- Announce that the first class to reach a designated number of points will be awarded a special privilege. (For example, thirty points, no homework for two nights in a row.) Display a point chart or use different colored marbles in a jar for each class so students can see at a glance the progress they are making toward their goal. Having students working toward a common goal can produce a positive effect on the entire class.

- Another effective way of encouraging problem students to behave is to use *peer pressure*. Set up the contingency that an entire class will earn a reward if a selected student (or students) improves his or her behavior. (For example: If Mike is on time and brings his materials, the entire class will earn one point. When the class has earned five points, they will receive five minutes of free time. Or: If the entire class hands in homework every day this week, there will be no assignment on Friday.)

Here are some positive reinforcers that a secondary class can work for:

- No homework one night
- Free time in class
- Time in class to do homework
- Conduct class on the lawn
- Class may listen to the radio for the final five minutes (they select station, you select volume)
- "Select-a-seat" day
- Popcorn while watching a film
- Break-time during class
- Open discussion
- Class trip
- Free homework pass redeemable in any class throughout the school (One teacher is assigned each week by principal to give out these passes)
- No homework over weekend
- Gift certificate for food

When determining positive reinforcement, don't forget the importance of supporting appropriate behavior daily. Though your secondary students may not openly seem to care about your praise and rewards, don't be fooled. Teen-agers, like all people, enjoy being acknowledged for a job well done. They just don't think it's "cool" to show it.

Remember, balance your negative consequences with frequent positive support.

Rewards
for Your Classroom

Once you have developed a system for recording positive classwide behavior, list the rewards you will use.

Individual Rewards

1.

2.

3.

4.

Classwide Rewards

1.

2.

3.

4.

Present Your Discipline Plan to Your Administrator

Your Discipline Plan, containing rules as well as negative and positive consequences, is now complete. Your next step is to share the plan with your administrator. Having administrative support will increase the effectiveness of your plan.

Share your Discipline Plan with your administrator and get his or her approval before you put the plan into effect.

Submit the plan in writing. Include the rules, the discipline hierarchy, and positive reinforcement ideas.

Discuss the administrator's role in your discipline hierarchy.

Discuss what the administrator will do when a student is sent from your class to his or her office.

Whenever you modify your plan, share the changes with your administrator.

Remember, you have a right to your administrator's support.

Present Your Discipline Plan to Your Class

When your plan is complete and has been approved by your administrator, present it to your class. Discuss the details of your plan, then display a Discipline Poster.

Write your rules, consequences and rewards on a large poster.

Laminate the poster so positives can be changed weekly and adjustments to your rules or consequences can be made.

Display the poster in a prominent location in your classroom. Be sure it is visible from the back of the room.

All students and visitors will then be aware of your discipline standards.

CLASSROOM RULES

1 _____
2 _____
3 _____
4 _____
5 _____

REWARDS

1 _____
2 _____
3 _____
4 _____
5 _____

CONSEQUENCES

1 _____
2 _____
3 _____
4 _____
5 _____
6 _____

LEE CANTER'S ASSERTIVE DISCIPLINE

When you meet students the first day, keep in mind that your classroom management begins at that moment. Make sure that you communicate the following expectations to them:

"I will tolerate no student stopping me from teaching or another student from learning."

"I will tolerate no student engaging in any behavior that is not in his or her best interest or in the best interest of others."

"You have a choice: Follow the rules and reap the benefits, or misbehave and accept the consequences."

Communicate Your
Discipline Plan to Parents

Once your discipline plan is completely developed and you have communicated your expectations to your class, send home a letter to parents. The letter should include your rules, positive and negative consequences, and a statement of your need for parental support. Include a tearoff portion that parents should sign and return to you.

Make this more pos.

Central Junior High School

645 E. Main Street
Columbus, Ohio 43113

Dear Parent(s);

    I will be your child's English teacher this year. In order to guarantee your child and all the students in my classroom the excellent educational climate they deserve, I have developed a Discipline Plan that will be in effect at all times.

    When in my classroom, students must comply with the following rules:

1. Follow directions the first time they are given
2. Be in seat, ready to work when the bell rings
3. Complete all assignments on time
4. Raise hand and wait to be recognized before speaking
5. Keep hands, feet and objects to yourself

If a student breaks a rule, the following consequences will occur:

First time student breaks rule: Name on board = warning

Second time student breaks rule: One checkmark, one detention

Third time student breaks rule: Two checkmarks, two detentions

Fourth time student breaks rule: Three checkmarks, two detentions, call parents

Fifth time student breaks rule: Four checkmarks, two detentions, call parents, send to principal

    If a student is severely disruptive, he or she will be sent immediately to the principal.

    Included in my Discipline Plan are ways to positively reinforce students who behave appropriately, In addition to using frequent praise, I will reward students with free class time, no-homework nights, and special class activities.

    In order for this plan to have its greatest effect, I need your support. Please discuss this letter with your child, then sign it and return it to me.

    Thank you for your cooperation.

    Sincerely,

- -

Parent/Guardian Signature   Date

Student's Name   Class

Comments:

Prepare a Discipline Plan
for Substitutes

To insure consistent discipline in your classroom, even when you are not present, prepare a Discipline Plan for substitutes. Make sure that a copy is left with the office. Put another copy in your plan book or tape it to the top of your desk.

From the desk of:

Dear Substitute:

The following are the guidelines for the Discipline Plan used in my classroom. Please follow them exactly, and leave me a list of students who broke the rules and a list of students who behaved properly. When I return, I will take appropriate action.

Class rules:

1. ....

2. ....

3. ....

4. ....

5. ....

The first time a student breaks a rule, record his or her name on the board. This constitutes a warning.

The second time a student breaks a rule, record one check next to his or her name. The consequence will be
....

The third time a student breaks a rule, record two checks. The consequence will be
....

The fourth time a student breaks a rule, record three checks. The consequence will be
....

The fifth time a student breaks a rule, record four checks. The consequence will be
....

If a student exhibits severe misbehavior such as fighting, open defiance, or vulgar language, he or she should immediately be sent to the principal.

Students who behave will be rewarded when I return with:
....
....
....

I appreciate your cooperation in following my Discipline Plan.

Sincerely,

Plan Action for Severe Behavior Problems

The greatest number of problems confronted by a teacher in the classroom are those students whose behavior is mischievous, disconcerting, disturbing and/or time-consuming. By consistently using the Assertive Discipline techniques just described, the majority of student misbehavior will be eliminated.

However, with some students who exhibit more severe problems, you may need to take stronger measures. Be aware that severe consequences should only be utilized when all else fails and that the guidelines should be strictly followed.

The following techniques have proved highly effective when used by secondary teachers:

- Use an Assertive Confrontation
- Develop a Behavioral Contract
- Tape Record Behavior
- Send the Student to Another Class
- Parent Attends Class
- Use a Discipline Card

> *Sending a student to a detention room or to an in-school suspension room are probably the most effective techniques for handling severe problems. Since these methods require total staff involvement they are discussed in detail in the next section, pages 46 to 48.*

Use an Assertive Confrontation

Problem students often act and feel as though they are the boss in the classroom. With such students it is often useful to set aside a specific time to sit down and openly confront their misperceptions and inappropriate behavior.

In an assertive confrontation, you communicate to the student the following message: "I am the boss in this classroom. There is no way I will tolerate your stopping me from teaching or someone else from learning. You will behave in my classroom!"

Here are some useful guidelines to follow when confronting a problem student.

Confront the student when you are calm and have planned what to say. Too often teachers sit down with a student when they are both upset.

Confront the student at a time when no other students are present.

Clearly and firmly tell the student exactly what you demand: "Do your work without talking back."

Tell the student what will occur if he or she does not comply: "I will immediately call your parents."

Tell the student why you are doing this: "I will not tolerate your disrupting my classroom. I care too much about you to allow you to behave this way in my classroom!"

To ensure that your demands are understood, have him or her repeat your statements to you.

If appropriate, have the principal and parents sit in during the confrontation.

Develop a Behavior Contract

Behavior Contracts are excellent aids in structuring interventions with problem students. A contract is an agreement between the teacher, principal, parent and student, and the contract is signed by all involved. (See page 37.)

Behavior Contracts should include the following:
- What you want the student to do (for example, come to class on time)
- What you will do if the student complies (provide points that will eventually earn the student the right to miss an assignment)
- What you will do if the student does not comply (the student will receive detention after school or during lunch)
- How long the contract will be in effect (1 week)
- Any contract should be designed so that the student can earn the positive consequence quickly.

Tape Record Behavior

The teacher places a cassette tape recorder next to the disruptive student. The recorder is turned on during the entire class period. Students will usually cease disrupting when the recorder is turned on, which is, of course, the goal of this technique.

The tape is played during a conference with the parent and principal. A plan for improving the student's behavior should then be discussed.

Send to Another Classroom

The disruptive student is sent to do his or her academic work in another classroom. For this consequence to be most effective:
- Plan your strategy with another teacher ahead of time.
- Send the student to a well-managed classroom in another curriculum area, i.e., English to Math or vice versa.
- The student should sit alone in back of class doing academic work.
- The student should not take part in any class activity while there.
- At the end of thirty minutes, the student is sent back to the regular classroom.
- Use this method as an alternative to sending a student to the principal's office.

Parent Attends Class

The parent comes to school and sits in on every class, including cafeteria and gym. To have its greatest effect, the parent must continue coming to school until the student shows improvement.

This method is successful because:

The parent sees for him or herself exactly how the child behaves in school and can take appropriate action at home.

The student feels pressure from peers about having his or her parent at school and begins to behave.

Discipline Card

When a student is a behavior problem for more than one teacher, the student needs to be assigned a Discipline Card.

All teachers involved with the problem student must meet and determine the rules, discipline hierarchy and positives that will appear on the card.

The student is assigned a Discipline Card before the school day starts and carries the card all day.

The first time the student disrupts in a classroom, the teacher initials the first consequence listed on the card; the second time, the second consequence, and so on.

Since the student receiving the card has been a persistent behavior problem, the first consequence on the card should *not* be a warning.

If the student loses the card, he or she receives all the consequences listed.

The student is assigned a Discipline Card as long as it is necessary.

The last teacher of the day sees that the student receives the negative or positive consequences.

A Discipline Card (see page 34) looks like this:

DISCIPLINE CARD
STUDENT _Susan Shaeffer_   DATE _11/12 - 11/16_

Behavior Teachers Expect:

Hand in all assignments on time.
Raise hand to speak.
Remain seated until bell rings.

CONSEQUENCES   INITIALS

1st _lunch detention_   BJ
2nd _15 minutes after school_   BJ
3rd _30 minutes after school, call parents_   KW
4th _On school suspension_
5th _Parent, principal, student conference_
Severe Clause _Send immediately to principal_

POSITIVES

If none of the consequences are initialed after one week, Susan is excused from one pop quiz.

Communicate With Parents
Throughout the Year

Your communication with parents should not be limited to twice-a-year routine parent-teacher conferences. To be most effective, you must establish positive communication with parents early in the school year. Then, when a problem that you cannot handle alone arises, contact the parents immediately. Do not wait until it is too late and the problem is out of hand. Use the following steps in developing a year-long plan for effective parent communication.

Communicate your standards to parents. Parents need to know your expectations if they are to support you. The first day of school send home a letter outlining your Discipline Plan. Have parents sign the letter and return it to you. (See sample, page 15.)

Positively reinforce students. Parents are accustomed to receiving only bad news from school. Sending home positive notes early in the school year will show parents you have a positive attitude towards their children. It will also increase your chances for gaining parental support if there is a problem.

Document all problems. Keep detailed records of inappropriate behavior. (See page 36.) Doing so will enable you to relate problems to parents in a fair, non-judgmental manner.

Be prompt. At the first sign of a problem, contact parents. Send a letter, place a phone call, or set up a conference the moment a problem arises. Do not wait until parent-teacher conferences or report card time. The problem will only get worse.

Good communication is vital to the success of your Discipline Plan. If you are having difficulty with a student and need to speak with a parent, do everything possible to reach him or her. Do not hesitate to call a parent at work. Though it may be difficult, you should not give up until you gain the parent's support.

Remember, the child is ultimately the parents' responsibility and you, the educator, deserve parental support.

Documentation

It is vital that teachers document all student misbehavior. This recordkeeping is necessary for:
- Student record cards
- Parent/teacher/principal conferences
- Transferring students to other classes
- Referring students for special counseling
- Placement of students in continuation or other special schools

There are two suggested ways of recording student misbehavior that secondary teachers have found effective.
- Use a Matrix System to record inappropriate behavior during the class period. It is an excellent substitute for the chalkboard. If your classroom rules are numbered, you can simply record the number, rather than the entire rule, on the matrix.

CLASS _Algebra_			DATE _12/2_		
Name	Rule Number 1	Rule Number 2	Rule Number 3	Rule Number 4	Rule Number 5
MARK NELSON		✓✓			
CAROL WHITE	✓✓✓				

- Later on, transfer the student's name and rule broken to a more permanent record. Also record the disciplinary action you have taken. Use a notebook or a section of a plan book with one page for each class period.

STUDENT	DATE	RULES BROKEN	CONSEQUENCES
Mark Nelson	12/2	Speaking out of turn	15 minutes lunch detention
Carol White	12/2	Did not follow directions	30 minutes lunch detention

Assertive Discipline
Analysis Worksheet

If your Assertive Discipline Plan is not working, you need to determine the reasons why. Read through this checklist to target your specific problems.

- ☐ You have not communicated your expectations to the class and parents.

- ☐ The students like the consequences.

- ☐ The consequences are not provided immediately.

- ☐ You are not consistent in providing consequences.

- ☐ Your plan is not tough enough.

- ☐ Your plan does not apply to all students.

- ☐ You only follow the plan occasionally.

- ☐ You are not using a classwide positive reinforcement program.

- ☐ You do not call parents.

- ☐ You do not keep detailed records of student misbehavior.

If you have checked any of the above reasons, review the first section, paying particular attention to the areas that you have checked.

Consistency is critical in eliminating discipline problems.

Sample Discipline Plan for General Classroom Use

The following is an example of a general classroom plan.

Rules

1. Follow directions the first time they are given.

2. Eating food and chewing gum is prohibited in class.

3. Come to class with all your materials.

4. Be in your assigned seat ready to work when the bell rings.

5. Raise hand and wait to be recognized before speaking.

Consequences

1st time student breaks rule	=	Name written down — warning
2nd time student breaks rule	= ✔	Lunch detention
3rd time student breaks rule	= ✔✔	30 minutes after-school detention, parents called
4th time student breaks rule	= ✔✔✔	After-school detention, parents called
5th time student breaks rule	= ✔✔✔✔	After-school detention, parents called, send to dean of students.
Severe clause	=	Send to dean of students

Positives

Free time at end of period

Opportunity to "throw" out lowest quiz grade

Time on computer

Play trivia game

The following is an example of a Discipline Plan for a shop class.

Rules

1. Follow directions the first time they are given.

2. Follow all safety rules when using classroom equipment.

3. All work areas must be cleaned before dismissal from class.

4. No equipment, tools or projects may be removed from class without the instructor's permission.

5. All assignments must be handed in on time.

Consequences

1st time rule is broken	=	15 minutes after-school detention
2nd time rule is broken	=	30 minutes after-school detention
3rd time rule is broken	=	In-school suspension, parents called
Severe clause	=	Send to principal

Positives

Early dismissal

Class may listen to radio

Free time in class

Classroom
Reproducibles

Chance Card

Distribute Chance Cards to reinforce appropriate behavior or good work habits.

Students write their names on the back and deposit them in a bowl or box.

At the end of the month, the names that are drawn receive a ''chance'' at rewards or small gifts that are raffled off.

Positively reinforce students by rewarding them with a Make the Grade Certificate for completing a specified number of homework assignments, having no tardies for a week, or for on-task behaviors. This certificate can be exchanged by the student for making the grade of "A" on one daily quiz or homework assignment.

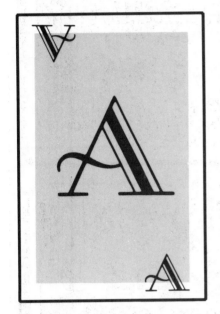

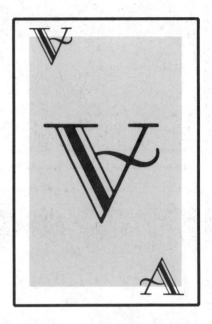

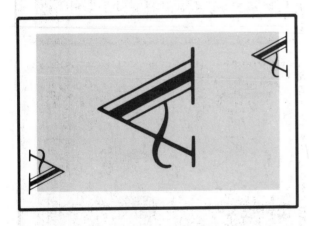
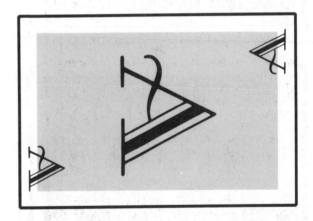

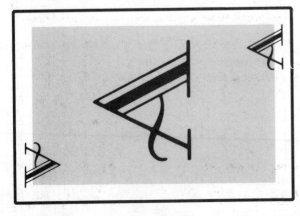
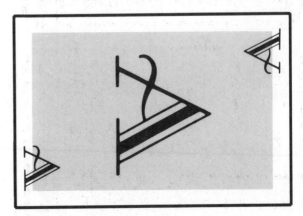

Walkman Rewards

Students who receive Walkman Rewards earn the privilege of using a radio with headphones during selected activities or at the end of the hour.

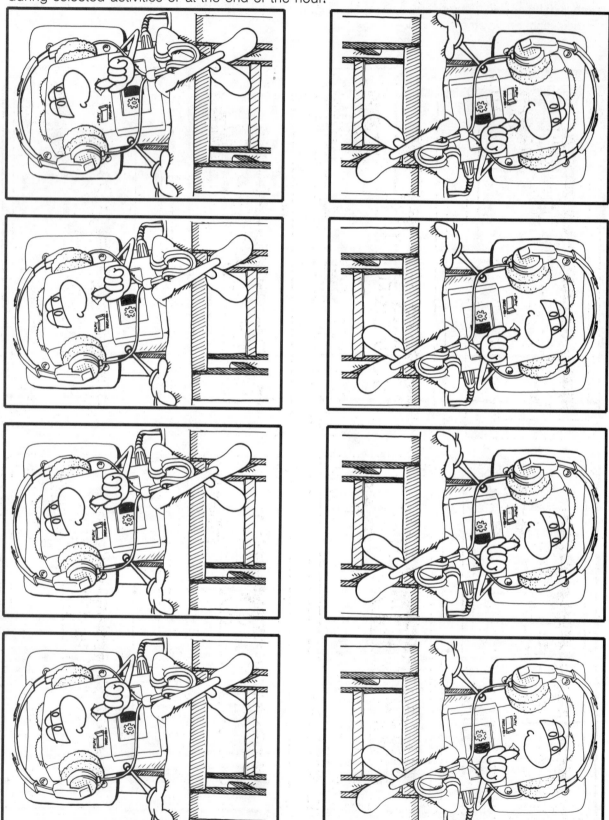

No-homework passes may be redeemed in place of any homework assignments in your class.

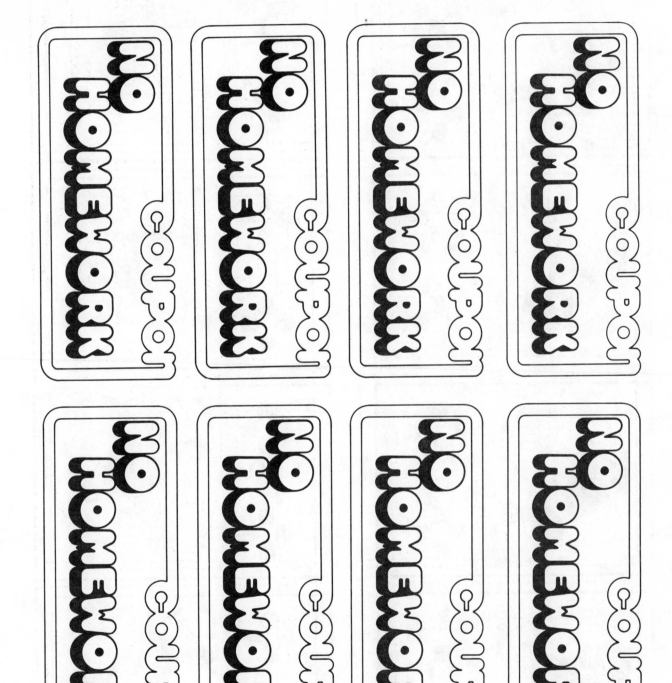

**SCHOOL OF EDUCATION
CURRICULUM LABORATORY
UM-DEARBORN**

Just Desserts

Students receiving these certificates may redeem them in the cafeteria on Friday for a special dessert.

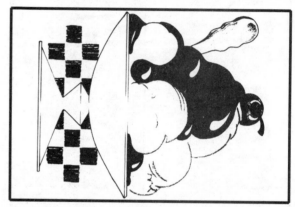

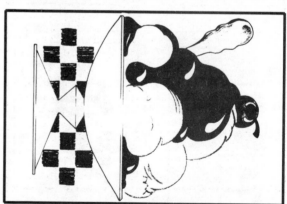

Popcorn Pass

Students may redeem these certificates for a cupful of popcorn during class movies or TV programs.

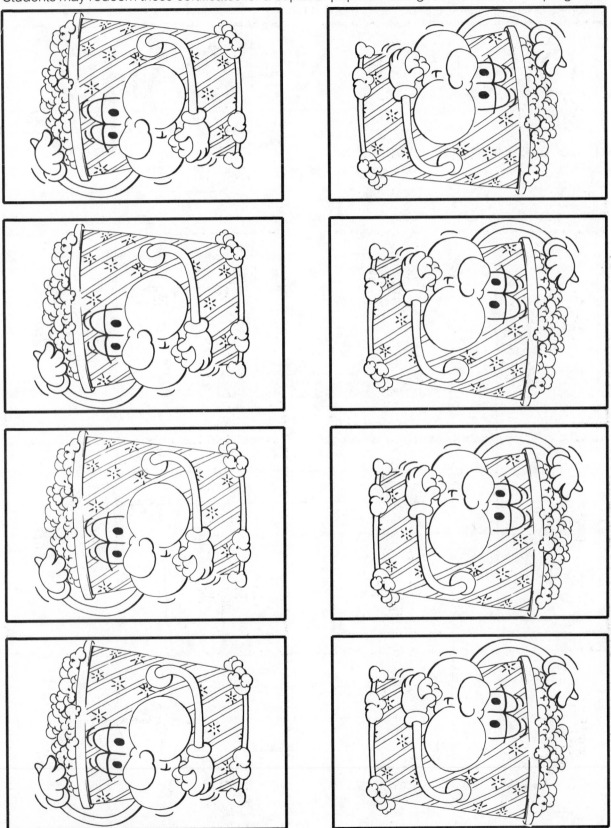

Discipline Card

Student carries card from class to class. Teachers initial consequences if student misbehaves. See page 20 for guidelines.

DISCIPLINE CARD

STUDENT_____ DATE_____

Behavior Teachers Expect:

_____

_____

_____

_____

CONSEQUENCES   INITIALS

1st _____   _____

2nd_____   _____

3rd _____   _____

4th _____   _____

5th _____   _____
Severe
Clause _____   _____

POSITIVES

_____

_____

_____

_____

_____

_____

_____

Matrix Documentation

Use this form to record misbehavior during classtime. It is an excellent substitute for the chalkboard. Whenever a rule is broken, place a check mark in the appropriate box. (See page 22.)

CLASS_____ DATE_____

Name	Rule Number 1	Rule Number 2	Rule Number 3	Rule Number 4	Rule Number 5

CLASS_____ DATE_____

Name	Rule Number 1	Rule Number 2	Rule Number 3	Rule Number 4	Rule Number 5

Discipline Record Sheet

At the end of a class period record the name, date, etc. of students who have misbehaved. This becomes your permanent documentation record. Reproduce one per class. (See page 22.)

STUDENT'S NAME	DATE	RULE BROKEN	CONSEQUENCES PROVIDED

Official Contract

Date _____

_____ promises to _____

_____ .

If student does as agreed, student will _____

_____ .

If student does not do as agreed, student will _____

_____ .

This contract will be in effect for _____ .

_____   _____
Student's Signature   Teacher's Signature

_____   _____
Parent's Signature   Principal's Signature

How to Develop a Schoolwide Assertive Discipline Plan

The goal of a schoolwide discipline plan is to have the entire staff develop a systematic, consistent way in which to deal with discipline problems throughout the school. To accomplish this goal there must be a two-sided effort from the staff:

1. Every teacher must have a classroom plan, i.e., a set of rules and consequences that are in effect at all times in his or her classroom. (See Section One.)

2. There must also be a schoolwide plan that governs student behavior in all of the common areas in the school.

Set Up a Committee

The development of a schoolwide plan should be the responsibility of a committee made up of representatives of the faculty, administration and department heads. Those chosen should be committed to making a positive change in the overall environment of the school.

Committee Responsibilities

The committee should follow the steps outlined in pages 42 to 53 to develop a rough draft of a schoolwide plan.

The rough draft should then be presented to the entire faculty for feedback and input.

The committee should use this information to write a final version.

Once the final version is approved, parents should be sent a letter from the principal explaining the schoolwide plan. (See page 53.)

Before it is put into effect, administrators should present the schoolwide plan to the entire student body.

Establish Schoolwide Rules

The committee should begin their task by discussing problems that occur in the common areas of the school. They should then formulate general rules designed to stop all the undesirable behaviors in these areas.

Here are two sets of examples of general schoolwide rules:

- Follow directions of staff.
- Stay in designated areas.
- No fighting.
- No dangerous objects, smoking or drugs.
- No vulgar language.

- Students will not litter anywhere on campus.
- Students will not call each other names, or use any vulgar language or signs.
- Students will not write on or damage the school's or another person's property.
- Students will not throw rocks or other harmful objects.
- Students will not leave the school campus or be in restricted areas during the day without permission.

The committee may feel that there are areas in the school that need specific rules (i.e., the hallway, cafeteria, yard, etc.). Here are two sets of examples of specific area rules:

Cafeteria

Follow directions of staff.
Students will not throw food.
Students must put all trash into the proper receptacles.
When waiting in line to be served, students must keep hands, feet
and objects to themselves.

Hallway

Follow directions of staff.
Do not run in the halls.
Always walk on the right side of the hallway.
Students must have a pass when in the hallways during class hours.

The committee may find it necessary to develop a separate tardy and/or truancy policy. Pertinent rules may be:

- Every student must be in class, in a seat, when the bell rings.
- Students will not miss a class without a written excuse from a parent.
- Students may not leave school without permission before the bell rings.

Establishing schoolwide rules will set standards throughout the school. Students will know that they are expected to behave in an appropriate manner both in the classroom and on the school grounds.

Schoolwide
Discipline Plan

List the general schoolwide rules:

1. _____
2. _____
3. _____
4. _____
5. _____

List rules for specific areas:

Hallway

_____

_____

_____

_____

Yard

_____

_____

_____

_____

Cafeteria

_____

_____

_____

Restrooms

_____

_____

_____

Campus

_____

_____

_____

Other

_____

_____

_____

Determine Disciplinary Consequences

Choosing Consequences

Once the committee has set the rules, they must decide how to deal with students who break the schoolwide rules.

As in the guidelines for the classroom plan, disciplinary consequences should be something the students do not like. The consequences should be arranged in order of severity, constituting a discipline hierarchy. For example:

1st time student breaks rule =   Warning
2nd time student breaks rule =   15 minutes detention, note sent home
3rd time student breaks rule =   30 minutes detention, note sent home
4th time student breaks rule =   1 day in-school suspension, parent called
5th time student breaks rule =   1 day in-school suspension, parent/principal/student conference

---

Parents should be given 24 hours notice if a student is to be detained after school or suspended.

There are many consequences that can be used as part of your hierarchy. **Here are some specific schoolwide consequences that secondary educators use:**
- Lunch detention
- Campus clean-up
- After-school detention (See page 46 for details.)
- Call parents
- In-school suspension (See page 48 for details.)
- Parent conference
- Send to assistant principal
- Saturday School (See page 49 for details.)

An immediate and more severe consequence (severe clause) should be determined for blatant misbehavior such as fighting, drug possession, defying authority, etc. **Severe consequences may be:**
- Immediately send student to principal and contact parents
- Placement in continuation school
- Recommendation for expulsion
- In-school suspension

Schoolwide
Disciplinary Consequences

List the consequences your school will provide for infractions of the schoolwide rules:

1st time _____

2nd time _____

3rd time _____

4th time _____

5th time _____

Severe Clause: _____

_____

List the consequences for tardies and truancies (if separate from general plan):

1st tardy/truancy _____

2nd tardy/truancy _____

3rd tardy/truancy _____

4th tardy/truancy _____

5th tardy/truancy _____

*The great majority of students want to be
in a safe, orderly, constructive environment.*

Organizing Consequences

Some of the consequences that the committee may choose to initiate will require discussion and preplanning. A room may need to be set aside, paperwork and forms developed, and faculty members may be asked to volunteer their services. Once the logistics for your particular school are agreed upon, write down the guidelines and post the schedules of staff responsibilities in the office. To help you set up specific schoolwide consequences, the following pages list some of the general guidelines you need to follow.

Detention Room

A detention room is any room designated as such for use before school, during lunch or after school.

The room is staffed by administrators and/or faculty on a rotating basis.

The disruptive student is assigned a specific amount of time in the room.

Students do academic work in the detention room.

Staff members may not talk with or counsel students while they are serving detention.

The staff member in charge is given a list of all students who are assigned detention and checks off the names of those who did or did not serve detention.

If a student does not appear, he or she receives double detention the next day or more severe consequences, i.e., in-school suspension.

If the student talks or disrupts in the detention room in any manner, he or she will be assigned extra detention and/or more severe consequences.

All teachers must volunteer time to supervise the detention room. Those who do not volunteer may not have the privilege of sending disruptive students there.

Instructions

1. Teacher fills out form, gives to student.
2. Teacher records student's name, date and time to be served.
3. Student gives form to supervisor in detention room. Supervisor fills out form and checks off student's name on master list.
4. Supervisor returns completed form to teacher.
5. Parents should be notified before student is to be detained after school.

DETENTION ROOM

STUDENT _John Henry_ DATE _5/15_

TEACHER _Ms. Saunders_ AMOUNT OF TIME _30 Minutes_

WORK TO BE DONE _History Assignment_

TIME IN	TIME OUT	
_12:00_	_12:30_	_Mr. Allen_
		SUPERVISOR'S SIGNATURE

COMMENTS:
Student completed assignment and followed detention room rules.

In-School Suspension

In-school suspension is the disciplinary action of removing a student from a scheduled class and placing him or her in an isolated, closely-supervised environment. Students attending in-school suspension are required to do work assigned by their regular teachers. In-school suspension is an alternative disciplinary action to placing students on out-of-school suspension. Since some students view spending a day at home preferable to being in school, especially if their parents are not home to supervise them, in-school suspension can be a more effective consequence. Use these guidelines when setting up an in-school suspension room:

The room is well-ventilated and well-lighted.

The room is monitored by an administrator, teacher, aide or other responsible adult.

The student does academic work in silence.

If the student disrupts in the in-school suspension room, he or she earns extra hours there.

The student eats lunch alone, and is escorted to and from the restroom.

The student remains in the room for a maximum of one day.

If the student misbehaves after returning to the classroom, he or she must return to the in-school suspension room.

Parents should be notified that students received in-school suspension.

The two most effective consequences for severe misbehavior are detention and in-school suspension. If your system is well organized and you follow the above guidelines, these two techniques will help to dramatically reduce discipline problems.

Send to Principal or Assistant Principal

This consequence will generally occur when the student has repeatedly disregarded the rules or has exhibited severe misbehavior. It is important that the principal determine ahead of time how a student will be disciplined when sent to the office. The principal should use a discipline hierarchy for students who are sent to the office. For example:

At the end of each visit to the office, the student should be told what will occur if he or she chooses to return.

A separate card should be kept for each student who comes to the office. The record should contain:
- Student's name
- Date
- Why he or she was sent
- What action principal took
- What will occur if student returns

Such records will help the principal decide on the appropriate discipline action.

1st time sent to principal	= Discussion with student, parent conference, or in-school suspension.
2nd time sent to principal	= Parent conference and/or in-school suspension.
3rd time and all other times	= In-school suspension, Saturday School, out-of-school suspension

Severe Squad

This method is effective for removing a severely disruptive student from a classroom or any area of the school. Four or five faculty members volunteer to form a "severe squad." A system should be developed for notifying the squad (bells, intercom, message, etc.) when a teacher is having difficulty and needs to remove a student from the class. The teacher summons the "severe squad" for assistance by whatever system you have preplanned for such emergencies.

Saturday School

Saturday School is an alternative to in-school or out-of-school suspension. It is a way of providing students with a severe consequence without causing them to miss valuable instruction time.

Teachers, administrators, staff personnel, or community members may volunteer to supervise Saturday School.

The hours are generally 8:00 A.M. to 12:00 P.M.

Students may be assigned campus clean-up or study time.

All of these procedures must be written into the school or district policy. Everyone in the school community must be notified of the procedures, including students, certified and classified staff and parents.

Monitoring Student Behavior

Having decided the consequences, the committee should discuss a system for monitoring student misbehavior in all of the common areas in the school.

Any adult may record the names of students who break the rules. This can be done by handing the student a pink slip or citation (a copy is sent to the office). The slip of paper should include the student's name, date, rule broken, and location in the school of the infraction of a rule. (See sample, page 63.) During the last period of the day, office personnel should go through the slips and indicate on a master list the consequences each disruptive student will receive. A computer is a useful tool for keeping track of pink slips. Consequences should be carried out the same day, or as soon as possible thereafter. Consequences should be cumulative over a marking period. That is, if a student disrupts three times in a marking period, that student receives the third consequence on the hierarchy, *not* three warnings.

Linda Forman
NAME

3/18
DATE

Littering
RULE BROKEN

Campus
LOCATION

B. Smith
SIGNED

Remember, every time a student breaks a rule, he or she should be provided a disciplinary consequence.

The monitoring system and disciplinary consequences you choose should be tailored to the needs of your school.

For example: In a small school a pink-slip system and discipline hierarchy may not be necessary. It may work well to send a disruptive student immediately to the office to be disciplined by the vice-principal.

On the other hand, in a large school with severe problems it may be necessary to utilize an elaborate monitoring system with frequent use of in-school suspension and Saturday School as consequences.

Remember, whatever system your school chooses, a cooperative staff that remains consistent can solve any problem, no matter how great.

Determine Positive Reinforcement

As in the classroom plan, negative consequences must be balanced with positive reinforcement in your schoolwide plan for the common areas of the school. A varied program of positive reinforcement will create a school that is an exciting place to be.

Choosing Positives

Your positive program should change frequently in order to continually motivate students to behave, and to continually reinforce those students who do behave appropriately. The committee can choose the consequences or a special **Positive Committee** made up of students can be formed. Use the following guidelines for the Positive Committee:

The committee should consist of students who deserve recognition for good behavior.

The committee members should change periodically.

One faculty member should sit on the committee.

The committee should choose the positives and run any raffles, etc., that are held.

The committee should select a special end-of-the-year reward and determine how students can achieve it.

Here are examples of some positive reinforcement ideas that students like to work toward:
- No-homework pass (redeemable in any class)
- Fast food coupons
- Discounts at local stores (i.e., record shop, hair stylist, drug store, computer store)
- Late pass/early dismissal pass
- Free lunch from cafeteria
- Letter home from principal
- Trips
- Films
- School dances
- Special luncheons

Monitoring Appropriate Behavior

A system must be developed for keeping track of and rewarding students who behave appropriately. The Gold Slip method is very effective.

Assigned staff members (especially those who patrol the halls and campus) may hand out Gold Slips as they "catch students being good." (See page 64.)   The adult simply fills in the student's name and date, then signs the slip and hands it to the student who is behaving in a commendable manner. The student deposits the slip in a container that is located in a central location (in or near the office, in the yard, etc.).

Try this: Select two or three faculty members each week who can distribute Blue Slips instead of Gold Slips. These Blue Slips are to be given to outstanding students and are worth 10 times the value of a Gold Slip.

The Positive Committee will decide what to do with the slips. **Here are some ways secondary schools reward students:**

Hold a raffle. Periodically (e.g., every 4 or 5 weeks) a raffle is conducted in front of the entire student body (or by grade level in a large school). A predetermined number of of Gold Slips are drawn and the winners receive any of the rewards previously mentioned.

Count the Gold Slips. Every week an assigned student or office personnel counts the slips and tallies the number of points by grade level. At the end of a designated period of time, the grade level with the most points receives a special reward (trip, film, luncheon, etc.).

Send home certificates. Particularly effective on the middle school level is sending home positive certificates for students who have earned a certain number of Gold Slips.

Remember, the student body should be notified ahead of time how the Gold Slips will be used to reward them. If they are working toward a goal, they need to know it. If they need a certain number of points, they need to know that too.

Try this: To encourage staff to be more positive, assign a quota. (e.g., Every faculty member must find 25 students doing something right this week!)

NAME _Dave Frank_

DATE _10/6_

COMMENDABLE BEHAVIOR _Followed rules_

LOCATION _Cafeteria_

SIGNED _Ms. Grant_

After a schoolwide plan has been formulated, a letter outlining the plan should be sent to parents from the principal. The letter should ask parents for their support.

Rockland High School

368 Towne Street   San Diego, California

Dear Parent(s) of Rockland Students:

The staff and I believe that our objective at Rockland is to educate children. We believe that in order for your child to meet the challenges that he or she will face in our society, the development of self-discipline and individual responsibility are essential. Over the past three years Rockland School has followed a successful Assertive Discipline Plan that supports the right of each student to an education in a calm, safe, and secure environment.

It is important that parents, school staff and students work together to maintain a positive educational atmosphere. The rules and discipline procedures at Rockland have been established for the protection of students. Students are expected to respect these rules as well as the people responsible for carrying them out. The staff needs the support of parents in promoting acceptable behavior. Our goal is for each student to learn to be responsible for his or her own actions.

The following serious violations will result in disciplinary action and may result in suspension:
1. Fighting
2. Failure to submit to the authority of adults
3. Destruction of property or theft
4. Possession of dangerous objects, drugs or alcohol
5. Habitual profanity

In addition, Rockland High School also has rules regarding the following:
1. Tardies
2. Truancy

Each classroom teacher has a similar set of rules and consequences that we can furnish upon request.

Students who do not follow the rules will be provided an appropriate consequence such as detention, in-school suspension, calls to parents. Well-behaved students will be recognized with group and individual rewards. We feel strongly that those students need to know we appreciate them and how much they contribute to our teaching/learning atmosphere.

Please read and discuss this discipline plan with your child. Sign and return the bottom portion of this letter. If you have any questions, please contact me and I will gladly discuss them with you.

With your continued cooperation and support, the school year will be a positive and motivating experience for the students are Rockland High School. Let's work together for the benefit of the children for whom we are responsible.

Sincerely,

Principal

DISCIPLINE POLICIES AND PROCEDURES

I have read the Discipline Policies and Procedures of Rockland High School.

Parents Signature_____ Date_____

Comments: _____

_____

Principal's Responsibilities

The following are guidelines for the principal or vice-principal to use when implementing a Schoolwide Discipline Plan.

Take the lead in organizing a schoolwide plan.
- Set up a committee to write the plan.
- Send a letter to parents.
- Speak with the student body regarding the plan.
- Take the lead in modifying the plan as needed.

Supervise the implementation of the plan in the yard, cafeteria, halls, etc.
- Train noon-duty aides, teachers and other personnel in how to implement the plan.
- Monitor the yard to make sure the rules are enforced consistently.
- Make sure positive consequences are provided for students who behave.

Evaluate and approve the Discipline Plan of each teacher.
- See that each teacher has a Discipline Plan.
- Discuss with teachers what will occur if a student is sent to the office from the classroom.

Establish a hierarchy of consequences for students sent to the office for disciplinary reasons. (See page 48.)

Establish positive consequences to be provided for students for appropriate behavior and/or improved behavior.
  For Example:
- Principal's award
- Letter to parents
- Special privilege

Provide back-up support for staff members who are working with students with severe behavior problems.
- Provide consequences whenever a student is sent to the office for breaking class or schoolwide rules.
- Set up a plan to assist teacher in removing severely disruptive students from the classrooms. (See Severe Squad, page 49.)
- Assist teachers in establishing specialized behavior contracts for special problem students.
- Be available to support teachers who are dealing with difficult parents of problem students.

Supervise your teachers' implementation of the Discipline Plan guidelines by observing teachers in their classrooms and giving them positive and negative feedback in relation to these key areas.
- Does the teacher clearly tell students what he or she wants?
- Does the teacher provide positive reinforcement consistently?
- Does the teacher provide a negative consequence — name on board, etc. — every time a student disrupts or does not follow directions?
- If the teacher's plan is not effective, does he or she modify it to make it work better?

Positively support your teachers.
- Teachers need positive support, too. Administrators, department heads, members of the Positive Committee should recognize teachers who are consistent and successful in their discipline efforts. Positive support can be in the form of:
  — Positive notes and letters
  — Free periods
  — Excused yard or cafeteria duty
  — Small gifts (coffee mug, flowers, pen, book)

At each staff meeting, discuss discipline problems and how they can be improved.

Dear Parents:

I am pleased to tell you that your son's behavior has improved tremendously. For two weeks now, he has done all of his classwork and has not received one pink slip on the campus.

I appreciate how cooperative you have been in disciplining him at home. Thank you for your support.

Sincerely,

Principal

Teachers' Responsibilities

The following are guidelines for teachers to use when implementing a Schoolwide Discipline Plan.

- Develop and follow through with a classroom plan.
  - Tell students what behaviors are expected of them.
  - Consistently reinforce appropriate behavior.
  - Provide negative consequences every time students disrupt.
  - Make sure the principal is aware of the plan.

- Provide input on guidelines for the schoolwide plan.

- Follow the guidelines of the schoolwide plan.

- Whenever necessary, assist fellow teachers, aides, etc., in dealing with behavior problems. For example, if a teacher observes a student breaking a rule in the yard, he or she should discipline the student even if the teacher is not on duty.

Sample Schoolwide Discipline Plan

The following is an example of a complete schoolwide plan.

I. Rules
1. Follow directions.
2. Stay in designated areas.
3. No littering on campus or in halls; no food in halls.
4. No provoking.
5. No vulgar language.

II. Consequences — Students who do not follow the rules will be issued pink slips.
1 pink slip   = Warning
2 pink slips = Lunch detention
3 pink slips = Lunch detention and after-school clean-up
4 pink slips = Call parents and after school clean-up

Severe clause = Referral to assistant principal

Any student using alcohol, drugs, dangerous objects, stealing or engaging in physical assaults against other students or staff will be recommended for explusion and/or charges will be filed with the local police.

III. Positives
Students who are behaving appropriately will be awarded Gold Slips.

Every six weeks there will be a drawing for each grade level. Students whose names are drawn will be awarded small gifts and certificates redeemable at local businesses (e.g., fast food, hair salons, sporting goods stores). Outstanding students will be recognized by special awards and letters to their parents from the principal.

IV. Truancy/Cut Policy
1st truancy/cut = Campus clean-up equal to the amount of time cut, parents called.
2nd truancy/cut = Campus clean-up equal to two times the amount of time cut, parents called.
3rd truancy/cut = Campus clean-up equal to three times the amount of time cut or Saturday school, parents called.

V. Classroom Plans
A. All teachers will have a classroom Discipline Plan.
   1. Rules
   2. Disciplinary consequences
   3. Positives
B. Students will be informed of these plans the first day of class.

VI. Assistant Principal Plan
A. Consequences
When a student is sent to the assistant principal for breaking rules anywhere in the school the following consequences apply:
   1st time = Conference and after-school detention
   2nd time = Parent conference and after-school detention
   3rd time = Parent conference, in-school suspension and program reevaluation.

B. Positive Consequences
Students who improve their behavior will receive a letter of commendation from the assistant principal.

Schoolwide
Reproducibles

DETENTION ROOM

STUDENT_____ DATE_____

TEACHER_____AMOUNT OF TIME_____

WORK TO BE DONE _____

_____

TIME IN	TIME OUT	
		SUPERVISOR'S SIGNATURE_____

COMMENTS:

DETENTION ROOM

STUDENT_____ DATE_____

TEACHER_____AMOUNT OF TIME_____

WORK TO BE DONE _____

TIME IN	TIME OUT	
		SUPERVISOR'S SIGNATURE_____

COMMENTS:

Notice to Parents

NOTICE OF ASSIGNED AFTER-SCHOOL DETENTION

To the
parent(s) of_____ :

Your son/daughter has been assigned _____ minutes after-school detention in

Room _____ with _____ on _____ .

The reason for this consequence is _____

chose to be disruptive in the following manner: _____

_____

_____

_____

_____

Please follow through at home regarding this matter.

_____   _____
SIGNED   DATE

NOTICE OF ASSIGNED AFTER-SCHOOL DETENTION

To the
parent(s) of_____ :

Your son/daughter has been assigned _____ minutes after-school detention in

Room _____ with _____ on _____ .

The reason for this consequence is _____

chose to be disruptive in the following manner: _____

_____

_____

_____

_____

Please follow through at home regarding this matter.

_____   _____
SIGNED   DATE

Pink Slips

Pink Slips are used to monitor misbehavior. See page 49 for explanation.

NAME DATE RULE BROKEN LOCATION SIGNED	NAME DATE RULE BROKEN LOCATION SIGNED
NAME DATE RULE BROKEN LOCATION SIGNED	NAME DATE RULE BROKEN LOCATION SIGNED
NAME DATE RULE BROKEN LOCATION SIGNED	NAME DATE RULE BROKEN LOCATION SIGNED

SCHOOLWIDE DISCIPLINE PLAN

Gold Slips

Gold Slips are used to monitor appropriate behavior. See page 52 for explanation.